内乡县衙打春牛

"人类非物质文化遗产代表作——二十四节气"科普丛书

春雨惊春清谷天
夏满芒夏暑相连
秋处露秋寒霜降
冬雪雪冬小大寒

中国农业博物馆 组编

Popular Science Book Series on
A Masterpiece of the Intangible Cultural Heritage of Humanity — The Twenty-Four Solar Terms

Whipping the Spring Ox at Neixiang County Yamen

Compiled and Edited by China Agricultural Museum

中国农业出版社
China Agriculture Press

北京 Beijing

丛书编委会
Editorial Board of the Book Series

主任 陈 通　刘新录
Directors

副主任 刘北桦　苑 荣
Deputy Directors

委员（按姓氏笔画排列 Ranking by Surname Strokes）
Members
王应德　王晓杰　王彩虹　毛建国　玉 云　皮贵怀
吴玉珍　吴德武　陈 宁　陈红琳　周晓庆　俞茂昊
唐志强　谢小军　巴莫曲布嫫

主编 苑 荣
Editor-in-Chief

副主编 唐志强　韵晓雁　王晓鸣　程晋仓
Deputy Editors-in-Chief

翻译 徐立新
Translator

本书编委会
Editorial Board of This Volume

主任 王晓杰
Director

副主任 王燕争
Deputy Director

主编 王晓杰
Editor-in-Chief

副主编 王燕争
Deputy Editor-in-Chief

参编人员 王 珊　李宗亚　王 晓　郭 涛
Editorial Staff

摄影 王晓杰　王燕争　苗叶茜　岳 立　王 勇
Photographers

"人类非物质文化遗产代表作——二十四节气"科普丛书
Popular Science Book Series on
A Masterpiece of the Intangible Cultural Heritage of Humanity — The Twenty-Four Solar Terms

序
Foreword

中华文化，博大精深，灿若星河，传承有序，绵延不绝。作为人类非物质文化遗产代表、凝结中华文明智慧的"二十四节气"在我国自创立以来，已经传承发展2 000多年。它是中国人观天察地、认知自然所创造发明出的时间知识体系，也是安排农业生产、协调农事活动的基本遵循，更是中国社会顺天应时、指导实践的生活制度。它是中华优秀传统文化中文明成果的典型代表，体现了传统农耕文明的智慧性，彰显了中国人认知宇宙和自然的独特性及其实践活动的丰富性，凸现了中国人与自然和谐相处的哲学思想、文化精神和智慧创造。

Chinese culture, wide-embracing and profound, brilliant in numerous fields, inherited in an orderly manner, has been developing without interruption. The "Twenty-Four Solar Terms" is a masterpiece of the intangible heritage of humanity and a crystallization of Chinese civilization and wisdom for 2,000 years of its existence. It is a time knowledge system invented by the Chinese people to observe heaven and earth and to learn about nature, the basic principles to organize agricultural production and coordinate farming activities, and also the life system for the Chinese society in its conformation to natural and meteorological timings and in its guidance of daily practices. As the typical representation of the fruit of the best of the traditional Chinese civilization, it embodies the wisdom of traditional farming civilization, reflects the Chinese people's unique interpretation of the universe and nature and the rich practices therein, and highlights their philosophical concepts, cultural spirit and intelligent creativity in their harmonious with nature.

"二十四节气"起源于战国时期，在公元前140年就已经有完整的"二十四节气"记载。从时间上，作为太阳历，早于儒略历（公元前45年）近一个世纪。"二十四节气"较之公历更准确地标识了地球视角的太阳运行规律。农谚就"二十四节气"同公历的关系说道："上半年是六廿一，下半年来八廿三。每月两节日期定，最多不差一两天。"这里所说的"不差"，不是"二十四节气"不准，而是公历有"差"。我们的生活要顺天应时，生活在自然体系之中，就应该把自己看成是包括自然界在内的客观世界的组成部分。无限制地扩大人的能力，破坏自然规律，其后果是难以意料的。

"人类非物质文化遗产代表作——二十四节气"科普丛书
Popular Science Book Series on
A Masterpiece of the Intangible Cultural Heritage of Humanity — The Twenty-Four Solar Terms

The "Twenty-Four Solar Terms" originated in the Warring States Period. There were already complete records of the "Twenty-Four Solar Terms" in 140 B.C., almost one century earlier than the Gregorian calendar (45 B.C.), also a solar calendar. It is more accurate than the latter in indicating the laws of the sun's movement from the earth's perspective. Agricultural proverbs identify its relationship with the Gregorian calendar: "In the first half of the year, the solar terms fall on the 6th and 21st of each month, and in the second half, they fall on the 8th and 23rd. There are two solar terms in each month, with an adjustment of one or two days." Here, "adjustment" is not the result of inaccuracy of the "Twenty-Four Solar Terms" but errors on the part of the Gregorian calendar. As we need to conform to the natural and meteorological laws and live in natural systems, we should regard ourselves as components of the objective world, including the nature. Expanding human capacity without constraint and disrupting the natural laws may lead to unexpected consequences.

对中国人来说,"二十四节气"是我们时间制度整体的一部分,它是指导我们包括农业在内的创造生活资料的一切活动的时间节律。而我们的情感表达、礼仪等调节人际关系、社会关系的活动则以对月亮运行周期观察为基础的太阴历为节律。我们的传统节日体系大都是以太阴历为依据的。在我们的阴阳合历的整体框架里认识"二十四节气",领会我们的先辈以置闰的办法精妙恰当地协调二者的对应关系,体现了中华传统文化的精奥和人文精神。

For the Chinese, the "Twenty-Four Solar Terms" is part of our time regime, the time prosody that directs all activities that produce living materials, including agriculture. The time prosody of activities that accommodate interpersonal and social relations, such as our emotional expression and etiquette and protocol, is the lunar calendar based on observations of the moon's periodic movement. Our system of traditional festivals is mostly founded on the lunar calendar. It is advisable to interpret the "Twenty-Four Solar Terms" in the overall framework of the lunisolar calendar, and understand how our ancestors wisely and aptly coordinate the correspondence between the solar and lunar calendars by means of intercalation, which reflects the beauty and humanistic spirit of traditional Chinese culture.

"二十四节气"融合四季，贯穿全年，广为实践，流布全国，影响世界。其作为我国优秀传统文化的典型代表和人类非物质文化遗产代表作项目，富含中国人特有的哲学思想、思维理念和人文精神，具有广泛的参与度和社会影响力，引发世人的关注与探索。2016年11月30日，在文化部非物质文化遗产司指导下，在中国民俗学会支持下，由中国农业博物馆作为牵头单位，联合相关社区单位申报的"二十四节气——中国人通过观察太阳周年运动而形成的时间知识体系及其实践"，被联合国教科文组织列入人类非物质文化遗产代表作名录。这是中国非遗保护工作取得的一项重要成果，也是对外文化交流的一次成功实践。在其带动影响下，全国人民乃至世界人民对"二十四节气"的认知、认同、参与和实践空前提高，进一步彰显和增强了中国人的文化自觉和文化自信。

The "Twenty-Four Solar Terms", integrating the four seasons and covering the whole year, is widely practiced throughout China, with influences on the whole world. As China's best representative of traditional culture and a masterpiece of the intangible heritage of humanity, it is full of philosophical thoughts, thinking patterns and humanistic spirit unique to the Chinese, enjoying a wide participation and social influence, and commanding attention and exploration from around the world. On Nov. 30, 2016, under the guidance of the Intangible Cultural Heritage Department of the Ministry of Culture and with the support of the China Folklore Society, "the Twenty-Four Solar Terms, knowledge of time and practices developed in China through observation of the sun's annual motion" submitted by the China Agricultural Museum together with related community organizations, was entered onto the list of Masterpieces of the Intangible Cultural Heritage of Humanity by the UNESCO. This was a significant achievement in China's intangible heritage protection, and also a successful practice in cultural exchange. Due to this endeavor and its influence, the people of China and of the world have unprecedentedly heightened their knowledge, identification, participation and practice regarding the "Twenty-Four Solar Terms", which further reflects and enhances Chinese people's cultural awareness and self-confidence.

出版这套"人类非物质文化遗产代表作——二十四节气"科普丛书，有助于在更大更广的范围和层面普及传播节气的相关知识，进一步增强遗产实践社区和群众的自豪感与凝聚力，激发传承保护的自觉性和积极性，扩大关于传统时间知识体系的国际交流与对话，推动人类文明交流互鉴。

The publication of this book series on *A Masterpiece of the Intangible Cultural Heritage of Humanity—The Twenty-Four Solar Terms* shall be conducive to its spread and popularization on a larger scale and in a wider sphere, further enhance the sense of pride and solidarity on the part of the inheritance practice communities and masses, inspire their awareness and initiative in preservation and protection, expand international exchanges and dialogues on traditional time knowledge systems, and promote exchanges and mutual learning between human civilizations.

期望并相信这套丛书能够得到社会各界人士的喜爱。

We sincerely hope and cordially believe that this series will win the hearts of readers of various circles.

谨为序。

Please enjoy your reading of this volume.

刘魁立
Liu Kuili

2019 年 3 月
March 2019

内乡县衙打春牛
Whipping the Spring Ox at Neixiang County Yamen

前 言
Preface

立春意味着春天的开始。民间又习惯地称立春为"打春",这与封建社会官办岁事活动"鞭春牛"有直接关系。"鞭春牛"又称"祭春牛",俗称"打春牛"。内乡县衙位于豫西南内乡县,距今已有700多年历史,是我国目前保存相对完整的封建时期县级官署衙门。旧时每年立春之际,知县都会率众官身着公服,高擎仪仗和春字牌到城南先农坛,祭祀春牛和芒神。知县依例鞭春牛并亲自扶犁耕地一垅,表示代御亲耕,劝民农桑,以迎春气而兆丰年。

The solar term "Beginning of Spring" is literally the start of the spring season. Traditionally it has been popularly known as "Whipping the Spring", and this originated directly from the annual ceremonial event "Whipping the Spring Ox" hosted by the government in the feudal society. The event was also known as "Giving Offerings in the Spring Ox", otherwise known as "Whipping the Spring Ox", The Neixiang County Yamen, located in the Neixiang County in the southwest of Henan Province and having a history of more than 700 years, is, by comparison, the best preserved county yamen – government offices at the county level – from the feudal periods in China. At the Beginning of Spring in old times, the county magistrate would be dressed in the official robe, and march with his large number of subordinates all the way to the Temple of Agriculture, with signs bearing his rank and signs of the character "spring" held high by his guard of honor, to offer sacrifices to the spring ox and the God of Spring. The county magistrate routinely whipped the spring ox and walked behind the ox with the plough along one ridge in a plot of field, signifying that he was farming in the place of the emperor, thus encouraging the people to be engaged in agriculture and silkworm cultivation, in an effort to embrace the breath of spring and to facilitate a bumper harvest.

2016年11月30日，联合国教科文组织将"二十四节气"列入人类非物质文化遗产代表作名录，"内乡县衙打春牛"作为十个非遗传承保护社区之一，被列入扩展名录。

On November 30, 2016, the UNESCO proclaimed the "Twenty-Four Solar Terms" a Masterpiece of Intangible Cultural Heritage of Humanity, and the "Whipping the Spring Ox at Neixiang County Yamen" was entered into the extended list of ten intangible heritage preservation communities.

目录 Contents

序 Foreword	
前言 Preface	
人文地理 A Cultural and Geographical Overview	1
历史渊源 Historical Origins	15
主要内容 Major Rites	24
相关设施 Existing Artifacts and Facilities	32
传承保护 Preservation and Protection	41

人文地理

A Cultural and Geographical Overview

内乡打春牛节气习俗的传承地——河南省南阳市内乡县，位于豫西南的南阳盆地西缘。东接镇平县，南连邓州市，西邻淅川县、西峡县，北依嵩县、南召县。境内山地面积占72.2%。北部山势呈西北—东南走向，中部和南部浅山南北延伸。县境属长江汉水流域，共有大小河流40余条。其中较大的有湍河、默河、刁河、黄水河、螺蛳河等。境内有天心洞、狄青洞、黄龙洞、天仙洞等知名的溶洞。

内乡县衙打春牛
Whipping the Spring Ox at Neixiang County Yamen

Neixiang County of Nanyang Municipality in Henan Province is located on the western rim of the Nanyang Basin in the southwest of the province. It neighbors on Zhenping County on the east, Dengzhou City on the south, Xichuan and Xixia Counties on the west, and Song County and Nanzhao County on the north. The total area of hilly regions within Neixiang County occupies 72.2% of its overall land space. The mountains in its north run from the northwest to the southeast, while the lower hills in the central and southern regions extend from the north to the south. The territory of the county falls within the basins of the Hanshui River and the Yangtze River, with 40 rivers of different lengthes running through. Major rivers include the Tuanhe River, Mohe River, Diaohe River, Huangshui River and Luoshi River. The county also has such famous karst caves as Tianxin Cave, Diqing Cave, Huanglong Cave, and Tianxian Cave.

内乡县湍河美景
Beautiful Scene of Tuan River in Neixiang County

内乡县衙打春牛
Whipping the Spring Ox at Neixiang County Yamen

　　内乡县资源丰富，物产富饶，森林覆盖率45.9%。大理石、花岗岩、米黄玉、海泡石、石墨、金、银、钒等22种矿产储量丰富；境内盛产石斛、麝香、天麻、何首乌、杜仲、辛夷、山茱肉、柴胡等400多种知名药材；内乡县还是全省优质烟叶和产粮基地。

Neixiang County has abundant natural resources and produces, with a forestry coverage of 45.9%. It enjoys a rich reserve of 22 minerals, including marble, granite, pale brown jade (yellow jade), sepiolite, graphite, gold, silver and vanadium. The production of over 400 famous medical herbs is abundant, including dendrobe, muskiness, tianma (Tall Gastrodia Tuber), heshouwu (Tuber Fleeceflower Root), tu-chung (bark of eucommia), violet magnolia (magnolia liliflora), Common Macrocarpium Fruit, and root of Chinese thorowax. It is also the provincial center for growing quality tobacco leaves and crops.

　　内乡县古属郦地，据旧志记载，西周时建郦国，春秋为楚郦邑，为楚国西北边境两个重镇之一。战国时期郦被秦占有。秦昭襄王三十五年，初置南阳郡，辖有郦县。西汉时期南阳郡辖包括郦县在内的36县。隋朝重建郡县，因境内有菊水，故郦县易名为菊潭。同时因避文帝之父杨忠之讳将中乡县改为内乡县，这就是内乡县名由来之始。

Neixiang County is part of what the old State of Li used to be. According to historical records, the State of Li was established in the Western Zhou Dynasty. It was Liyi (City of Li) in the State of Chu in the Spring and Autumn Period, one of the two major strategic towns in its northwest. During the Warring States Period, it was occupied by the State of Qin. In the 35th year during the reign of King Zhaoxiang of the State of Qin, Nanyang Jun (or prefecture) was first designated, with jurisdiction over the territory of the old State of Li. In the Western Han Dynasty, Nanyang Jun had 36 counties under its jurisdiction, including the County of Li. In the Sui Dynasty, the County of Li was reestablished, and due to the Ju River in its territory, it was renamed

County of Jutan (Ju pool). Meanwhile, to avoid overlapping with any part of the name of the father of Emperor Wen, Yang Zhong, Zhongxiang County was renamed Neixiang County, which was the origin of the name of the County.

内乡县历史悠久，人杰地灵，曾培育和造就了明代政治家柴升、清代理学家王检心等历史文化名人。

Neixiang County has a long history, glorious for the many famous people who have lived here. It has produced such highly regarded personages in history and culture as Ming Dynasty politician Chai Sheng and Qing Dynasty Neo-Confucianist Wang Jianxin.

内乡县旅游资源得天独厚，拥有被联合国教科文组织列入世界生物圈保护区的国家AAAA级景区——南阳伏牛山世界地质公园组成部分的内乡宝天曼生态旅游区；中原生态养生福地——云露山；商圣范蠡的纪念祠堂——商圣苑以及宝天曼峡谷漂流、七星潭景区、桃花源景区、天心洞等众多风景名胜。人文景观有中国景观村落——吴垭石头村；有中国目前保存最完整的封建时代县级官署衙门——内乡县衙（国家级重点文物保护单位）；河南省目前保存较为完整的古代文庙建筑之一的内乡文庙；新石器时期的茶庵遗址、小河遗址、朱岗遗址等；宋代北方汝窑系瓷窑——邓窑遗址；民国时期的建筑史迹——内乡湍河老桥；始建于清乾隆年间，距今已有260余年历史的吴垭民居村；始建于明代的法云寺塔以及省立信阳师范学校旧址等，均为省级文物保护单位。还有42处县级文物保护单位。

The County has very unique tourist resources, including Neixiang Baotianman Ecological and Cultural Tourism Area as part of the Nanyang Mount Funiushan

内乡县衙打春牛
Whipping the Spring Ox at Neixiang County Yamen

Global Geopark, a national 4A-level scenic area listed as a World Biosphere Reserve by UNESCO; Yunlu Mountain, an ecological health-preserving land of promise; Shangshengyuan, the memorial hall dedicated to the God of Wealth Fan Li; Baotianman Canyon Drifting; Qixingtan Scenic Area; Taohuayuan Scenic Area, Tianxindong Caves and a host of other scenic spots. Cultural landscapes include Wuya Shitou Village, the Chinese landscape village; Neixiang Confucian Temple, an ancient Confucian temple architecture currently preserved intact in Henan Province; Cha'an Relic, Xiaohe Relic, Zhugang Relic, from the Neolithic Age; the Dengyao Kiln Relic, a porcelain kiln of the Northern Ru Kiln System of the Song Dynasty; Neixiang Tuanhe Old Bridge, architectural relic of the Republic of China period; Wuya Ethnic Village, first established during the reign of Emperor Qianlong in the Qing Dynasty, with a history of over 260 years; Fayun Temple Pagoda, first established in the Ming Dynasty; and the former site of the Xinyang Normal School of Henan Province. All the foregoing are provincial-level cultural relic protection units. There are also 42 county-level cultural relic protection units.

内乡县非物质文化遗产丰富。明末清初陕西东路秦腔传入南阳后，与当地民歌小调融合而成的珍稀剧种——宛梆，距今约有400年历史。2006年，内乡宛梆被列入第一批国家级非物质文化遗产名录。《王莽撵刘秀》《宝天曼》《石堂山》等是内乡县流传至今的经典传说故事；深受众多专家学者赞誉的"马山童谣"被誉为豫西"儿歌经典"和"童谣至尊"的代表作；封建社会县级行政长官训谕和教化百姓的代表形式——宣讲圣谕，至今仍是当地一项重要的活动形式；此外，还有内乡县分布较广、流行较早的民间舞蹈——"竹马舞"以及民间文学艺术表演形式——鼓词，又称说书，俗称"鼓儿哼"等。

Neixiang County enjoys a rich intangible cultural heritage. After Qinqiang Opera of Shaanxi Eastern Route was introduced into Nanyang towards the end of the

Ming Dynasty and at the beginning of the Qing Dynasty, it was integrated with the local folk melodies, and evolved into a rare and treasured opera category: Wanbang Clapper Opera, with a history of 400 years. In 2006, Wanbang Clapper Opera was of the first group included in the list of Chinese National Intangible Cultural Heritage. "Wang Mang Chasing Liu Xiu", "Baotianman" and "Shitangshan" are all Wanbang Clapper Operas based on classic legends popular in the Neixiang County. "Mashan Nursery Rhymes" held in high esteem by many experts and scholars are reputed to be the representative works of "classic nursery rhymes" and "ace nursery rhymes"

《宣讲圣谕》节目演出
Performance of "Preaching the Imperial Decrees"

in western Henan Province. Preaching imperial decrees, as the most representative approach adopted by county magistrates in the feudal society to cultivate the mind of the masses, is still an important local custom. Furthermore, a folk dance called "Zhuma Dance" (bamboo horse dance) is widespread and has been popularized since very early on in Neixiang County. A literary performance, Guci (rhythmic chanting), is otherwise known as story-telling, popularly called "Gu'erheng" (humming to the tune of the drum).

宛梆表演
Wanbang Clapper Opera Performance

内乡县衙打春牛
Whipping the Spring Ox at Neixiang County Yamen

竹马舞
Zhuma Dance

鼓词
Guci (rhythmic drum chanting)

位于内乡县城,号称"天下第一衙"的内乡县衙,始建于元大德八年(1304年),历经元、明、清,距今已有700多年的历史,现存建筑多为清代建筑。1984年被批准为全国第一家衙门博物馆,是目前我国保存最完整的皇权时代县级官署衙门。县衙占地47 000米^2,有院落18进,房舍280余间,珍存文物近千件,其建筑和文物遗存从不同的侧面反映了我国皇权社会基层政权运作模式和官吏的生活起居,是一座研究皇权社会珍贵的"文史资料库"。可谓"一座内乡衙,半部官文化"。

内乡县衙宣化坊
Xuanhuafang (memorial gateway) of Neixiang County Yamen

内乡县衙打春牛
Whipping the Spring Ox at Neixiang County Yamen

The Neixiang County Yamen, known as "the No.1 county yamen under heaven" and located in the county seat of Neixiang, was first established in the eighth year of the Dade Era (1304), with a history of more than 700 years. It existed through the Yuan, Ming and Qing Dynasties, although the bulk of the existing architecture was first built in the Qing Dynasty. In 1984, it was approved to be the first yamen museum in the country, the existing county yamen left from imperial times that has been kept intact in China. The yamen occupies an area of 47,000m², with 18 depths of courtyards, more than 280 rooms, almost one thousand items of treasured cultural relics. The architectural and cultural relics reflect the operational model of grassroots authorities and daily life styles of officials of this stratum in the imperial society of China. It is reputed to be the valuable "cultural and historical databank" for studying the imperial Chinese society. People exclaim, "the Neixiang County Yamen can tell half the story of the bureaucratic culture".

三省堂
Samxing Hall (Hall of Three Reflections)

牛春打渐县乡内
Whipping the Spring Ox at Neixiang County Yamen

主簿衙
Administrative Office Building of Magistrate's Aide

县衙大门
Gate to the County Yamen

夫子院
Confucius Courtyard

戒石坊
Jieshifang (Admonition Memorial Stone Archway)

　　内乡县衙的建筑与《明史》《清会典》所载"坐北朝南、左文右武、前堂后邸、监狱居南"的建筑规制完全一致，其大堂、二堂、三堂分别比附北京故宫的太和、中和、保和三大殿而建，秉承我国古代皇权社会的中庸和礼制思想，同时又受其地理位置和其主持营建者、工部遣派官员浙江人章炳焘的影响，融南北建筑风格于一体，充分展示了我国古代劳动人民的建筑艺术，被专家称为"北有北京故宫，南有内乡县衙""龙头在北京，龙尾在内乡"。

内乡县衙打春牛
Whipping the Spring Ox at Neixiang County Yamen

内乡县衙二堂屏门匾——天理国法人情
"Heavenly Rules, National Laws, Human Affections"—Plaque on Screen Gateway of the Second Depth of Courtyard of Neixiang County Yamen

The architectural arrangements of the Neixiang County Yamen correspond perfectly with records of the architectural rules in *Ming Shi* (A History of Ming Dynasty) and *Qing Hui Dian* (A Record of Laws and Systems of Qing Dynasty): "The county administrative offices face south, the left-hand side of the hall reserved for civilian officials and the right-hand side kept for military officers, with the administrative halls in the front and residential quarters in the rear, and the prison set in the south". Its front hall, second hall and third hall were built in symbolization of the Hall of Supreme Harmony, the Hall of Central Harmony, and the Hall of Preserving Harmony of the Forbidden City. It follows the doctrine of the golden mean and the concept of social institutions and etiquettes. It was influenced by its geographic location and ideas of Zhang Bingtao, the managing architect and official assigned by the Ministry of Works, whose hometown was in Zhejiang. It thus integrated architectural styles of both north and south China, fully demonstrating the architectural art of the working people of ancient China. Experts claim, "The Forbidden City of Beijing is in the north, and the Neixiang County Yamen is in the south", "The head of the Chinese dragon is in Beijing while the tail of the dragon is in Neixiang".

内乡县衙打春牛
Whipping the Spring Ox at Neixiang County Yamen

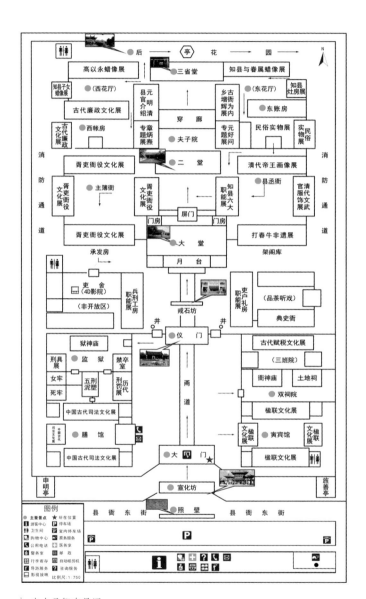

内乡县衙布局图
Layout of Neixiang County Yamen

历史渊源
Historical Origins

封建社会的经济主要是农业，农业经济占所有经济产业的80%以上，农业的丰歉直接关系黎民生活、国家财政收入和社会的安定，特别是闭关锁国政策后，清代知县的六大职能中的劝民农桑一职就更受重视。皇帝为了表示对农业生产的重视，每年立春之日，都要到先农坛祭祀先农，并扶犁亲耕。上行下效，在县级衙门，知县为表亲民、劝农，除了在立春前一天扶犁亲耕，还要在立春当天鞭打春牛，迎春气，与民同乐。

内乡县衙打春牛
Whipping the Spring Ox at Neixiang County Yamen

In the feudal society, the main economic sector was agriculture, which occupied over 80% of the economic sectors. The agricultural harvest had direct impact on people's livelihood, the national revenue, and social stability. This was especially so in the Qing Dynasty when the country was closed to the outside world, and thus encouraging the people to be engaged in agriculture and silkworm farming, as one of the six major functions of the county magistrates in the Qing Dynasty, was given even more priority. The emperor at that time would visit the Temple of Agriculture on the day of the Beginning of Spring to pay respect to Xiannong, the God of Agriculture, and plow the fields behind the ox as a gesture of giving priority to agriculture. This example was followed at local levels at the county yamen. In order to show his affinity with the people and encouragement of farming, apart from the gesture of plowing the fields on the day before the Beginning of Spring, the county magistrate would whip the spring ox and usher in the breath of spring on the day of Beginning of Spring, celebrating the festival together with the people.

内乡县衙打春牛
Whipping the Spring Ox at Neixiang County Yamen

"一年之计在于春"。"立春"为春季的第一个节气，民间习称"打春"，预示新的一年春暖花开、万物复苏。牛是人类的朋友，在十二生肖中是最勤劳的动物，在自给自足的自然经济社会里占有十分重要的地位。中国几千年的封建社会，农业一直是赖以生存的物质基础，历来受到统治阶级和普通百姓的高度重视。"打春日"和"牛"的结合，就有了封建社会官办岁事活动"打春牛"，也称"鞭春牛"。春季打春牛预示农耕活动的开始，有规劝农事、鼓励春耕的含义。

| 内乡县衙打春牛
Whipping Spring Ox at Neixiang County Yamen

内乡县衙打春牛
Whipping the Spring Ox at Neixiang County Yamen

"A year's plan starts with spring". Beginning of Spring is the first solar term in the spring season, popularly known as "whipping of spring", indicating that in the New Year, it is already warm and the flowers blossom and all life comes back to life. The ox is man's friend, and the most diligent animal among the 12 animal signs of the Chinese zodiac, playing a vital role in the society of self-reliant natural economy. In the Chinese feudal society of several thousand years, agriculture was always the material foundation for survival, given a high priority all along by the ruling class and the masses. From the combination of the "Day of Whipping the Spring" and the "ox" has evolved the annual ritual of "Whipping the Spring Ox". When everything comes to life in spring, "Whipping the Spring Ox" indicates the beginning of farming, signifying an encouragement and urging of agriculture and plowing in spring.

史料记载,"打春牛"这一活动,源于周代《礼记·月令》中的"出土牛,以送寒气"之说。土牛,即泥土所制的牛。唐代高承《事物纪原》记载:"周公始制立春土牛,盖出土牛以示农耕早晚"。到了汉代,打春牛风俗已相当流行。立春日清晨,京城百官身着青衣、戴青帽、立青幡,送土牛于城门外,官员执鞭击土牛,以示劝农的迎春。这种仪式,已经固定下来,并传至各郡县。

According to historical records of *Li Ji · Yue Ling* (Proceedings of Government in the Different Months, Book of Rites), as an organized event, "whipping the spring ox" originated in the Zhou Dynasty: "An earthen ox was made in order to see off the wintry breath". According to *Shi Wu Ji Yuan* (The Records of Origin on Things and Affairs) by Gao Cheng in the Tang Dynasty, "Duke Zhou of the Zhou Dynasty started the custom of making earthen oxen and the making of such signified the timing of the start of agriculture and farming". By the Han Dynasty, the custom of whipping the spring ox had been very popular. On the morning of the Beginning of Spring, officials of the royal court in the capital put on greenish blue costumes and black hats and upheld greenish blue banners, walked behind the earthen ox to the outside of the city gate, where they whipped the earthen ox as a gesture to welcome the spring and to urge farming. This ceremony became customary and was passed on to the various *juns* and counties.

知县扶犁亲耕
County Magistrate Driving the Ox and working the plough to Plow the Field

　　唐代，打春牛活动相沿成习。唐代以前不太讲究土牛、芒神（即句芒、司春之神）的颜色，而到了唐代，则土牛、芒神的颜色则要"各随其方"，即一座城中，东西南北各造一土牛、芒神，其颜色必须与城门所在的方位相对应，即城东青色，城南赤色，城西白色，城北黑色。到了宋代，打春牛更加普遍。据《东京梦华录》记载："立春前一日，开封府进春牛入禁中鞭春"。宋仁宗颁布《土牛经》后，鞭春牛之风日

内乡县衙打春牛
Whipping the Spring Ox at Neixiang County Yamen

益活跃，由宫廷、官署而遍及乡里，使鞭土牛风俗传播更广，以至成为民俗文化的重要内容。"小儿着鞭鞭土牛，学翁打春先打头"，杨万里《观小儿戏打春牛》一诗生动地记述了宋代"鞭春牛"活动。

In the Tang Dynasty, the rituals of whipping the spring ox had been developed into a customary practice. Before the Tang Dynasty, people did not care about the colors of the earthen ox and of the God of Spring, but in the Tang Dynasty, their colors had to "conform to their localities where they are situated". In other words, in each city, one earthen ox figure and one figure of the God of Spring were made respectively in the east, west, north and south of the city, and the their colors had to correspond to the direction where the city gate was located: greenish blue in the east, red in the south, white in the west, and black in the north. In the Song Dynasty, whipping the spring ox became even more widespread. According to *Dong Jing Meng Hua Lu* (The Eastern Capital: A Dream of Splendor), "one day before the Beginning of Spring, the magistrate of Kaifeng Prefecture, where the capital city Kaifeng was located, would bring a spring ox into the royal palace for the emperor to whip". After Emperor Renzong of the Song Dynasty issued the *Tu Niu Jing* (Rules on the Earthen Ox), the custom of whipping the spring ox became even more popular. It was spread from the royal court, to government offices and ultimately to all towns and villages, becoming a major item of local folk customs. "When a little child whips the ox, he can only do this wherever he can reach the earthen ox. When an old scholar official whips the ox, he does it on its head". This was a vivid depiction of the event of "Whipping the Spring Ox" in the Song Dynasty in poet Yang Wanli's piece "Watching Small Children Whipping the Spring Ox for Fun".

元代承袭了"鞭春牛"的风俗。宫廷中亦有迎春牛、鞭春牛等活动。每年立春前，太史院先要奏报立春具体日期，并且移文宛平县或大兴县，准备春牛、句芒神等。立春前三天，太史院、司农司请中书省宰辅等官员一同在大都齐政楼南迎接太岁神牛。立春当天清晨，"司农、守土正官率赤县属官具公服拜长官,以彩杖击牛三匝而退。土官大使送句芒神入祀。"(《析津志辑佚·岁纪》),中书省、户部向皇帝、太子、后妃、诸王、宰辅及各种中央官衙进送春牛。

The Yuan Dynasty inherited the custom of "whipping the spring ox" and in its royal court there were also the events of "Welcoming the Spring Ox" and "Whipping the Spring Ox". Before the Beginning of Spring each year, the Taishiyuan (Office of Astronomy and Observatories) submitted reports to the emperor regarding the specific date of Beginning of Spring and issued orders to the Wanping County or Daxing County government to prepare figures of the spring ox and the God of Spring. Three days before the Beginning of Spring, the Taishiyuan and the Ministry of Agriculture would invite the prime minister of the Imperial Secretariat and other officials to welcome the holy ox at Qizheng Plaza in Dadu (Beijing). On the very morning of the Beginning of Spring, "the Minister of Agriculture and provincial governors, together with petty officials at the county level, put on their official robes, kowtow to their superiors, whip the earthen ox with a colored stick three times and then take their leave. Provincial governors and other high-ranking officials then accompany the God of Spring into the royal court for rituals." (*Xi Jin Zhi Ji Yi* · *Sui Ji* <A Collection of Sporadic and Anecdotal Records of Xijin (Beijing) · A Year's Activities>). The Imperial Secretariat and the Ministry of Revenue would submit a spring ox respectively to the emperor, the crown prince, royal concubines, the prime minister and various central government administrations.

明代,迎春礼仪更加隆重而丰富。《明会典》中明确规定《有司鞭春仪》,除用土牛迎接春天的到来,以验占散病逐疫、祈求丰收、祈求吉利财旺之外,人们更加重视鞭春牛的习俗。认为鞭春牛之土,一可以涂灶却虫蚁,二可以宜桑宜畜,三可以得春为吉。

In the Ming Dynasty, the spring-welcoming rituals were even more ceremonious and colorful. *Ming Hui Dian* (Code of Great Ming Dynasty) had an exclusive

内乡县衙打春牛
Whipping the Spring Ox at Neixiang County Yamen

chapter with the heading of "Specifications for the ceremony of whipping the spring by various government offices". According to this, people welcomed the spring with the earthen ox in order to dispel diseases and quell epidemics and to pray for bumper harvests, best luck and great fortune. Meanwhile, people laid more emphasis on the custom of whipping the spring ox. They believed that dirt brought home from the earthen spring ox whipped could serve three functions: coating the kitchen stove to dispel insects and mosquitoes, facilitating mulberries and domestic animals, and serving to bring good luck for spring.

清代，鞭春牛活动蔚然成风。清潘荣升《帝京岁时纪胜》中记载："立春日，大兴宛平县令，设案于午门外正中，府县生员舁进，礼部官前导，尚书、侍郎、府展及丞后随，由午门中门入。"那轰轰烈烈的场面阵势，一点儿没有变。清末时期又改用纸牛，即以竹为骨架，外糊以纸。

In the Qing Dynasty, the custom of whipping the spring ox became amazingly pervasive. According to *Di Jing Sui Shi Ji Sheng* (A Record of Great Customs of the Capital City Beijing in the Four Seasons) by Pan Rongsheng, "on the day of the Beginning of Spring, the magistrates of Daxing and Wanping Counties set up tables in the very center of the square outside the Meridian Gate. County officials and students of government-sponsored academies carried the earthen ox here. Ministers, vice ministers, governors, mayors and their deputies followed behind, and entered through the middle gate of the Meridian Gate." The impressively enthusiastic spectacles had been preserved, without any loss of the original atmosphere and flavor. Towards the end of the Qing Dynasty, they started to use paper oxen with bamboo as the skeleton and wrapped up and pasted it in paper.

内乡县衙打春牛
Whipping the Spring Ox at Neixiang County Yamen

芒神牵牛
God of Mang Leading the Ox

知县打春
County Magistrate Whipping the Spring Ox

主要内容
Major Rites

　　作为保存最完整的封建时期县级官署衙门，内乡县衙在每年立春之时，也依例"鞭春牛"，以示春意兆丰年。清同治八年《内乡通考》记载了同治年间迎春及芒神春牛的活动，不仅证明了鞭春习俗在清代内乡县衙的沿袭，也说明这些费用是由官方支付，显示出县衙对于迎春鞭春的重视。据上述史料记载及内乡城中老人回忆，县衙礼房负责按照官方尺寸制作春牛，每年六月由掌管天象、历法的朝廷官员钦天监预定次

内乡县衙打春牛
Whipping the Spring Ox at Neixiang County Yamen

年春牛、芒神的造型、颜色，命令各府州县以式制作。其春牛和芒神尺寸都有规定和象征意义。内乡县衙的石制迎春池也在鞭春活动中起到了重要作用，平时贮备的雨水称为无根之水，在制作春牛时使用，有迎春、敬春的美好寓意。

As the county yamen in feudal times preserved most intact in the country, the Neixiang County yamen followed the same precedent and custom of "whipping the spring ox" on the Beginning of Spring each year to signify the thickness of spring ushering in a year of bumper harvests. *Nei Xiang Tong Kao* (A Complete Survey of Documents and Regulations of the Neixiang County), compiled in the eighth year during the reign of Emperor Tongzhi of the Qing Dynasty (1869), recorded the ritual events of welcoming the spring, the God of Spring and the spring ox during the reign of Emperor Tongzhi. This indicated that the custom of whipping the spring was passed down in the yamen of Neixiang County in the Qing Dynasty, and also indicated that these expenses were paid by government offices and that the county government gave emphasis to the welcoming and whipping of the spring. According to the abovementioned historical records and the recollections of old people in the city of Neixiang, the protocol department of the county yamen produced the spring ox according to officially specified measurements. Every June, royal court officials in charge of astronomical and calendar affairs at the Bureau of Astronomy placed orders for the styles, patterns and colors of the spring ox and the God of Spring figures, and commanded the provinces, prefectures and counties to make them for use in the following spring. There were specific requirements for the measurements of the figures to be made. The spring ox is four *chi* (one *chi* is 33.3cm) tall, symbolizing the four seasons of the year; it is eight *chi* long, signifying eight solar terms including the Beginning of Spring; its tail is one *chi* and two *cun* (one *cun* is 3.33cm) long, indicating the twelve months of the year. The God of Spring is three *chi* six *cun* and five *fen* (one *fen* is 0.333cm) tall, symbolizing the 365 days of the year; the whip in its hand is two chi and four cun long, representing the twenty-four solar terms of the year. The stone-made spring-welcoming pool at the Neixiang County Yamen also played an important role in the spring-whipping rituals. The rainwater saved from rainfalls was called rootless water, to be used in the making of the spring ox, with an auspicious significance of welcoming and revering the spring.

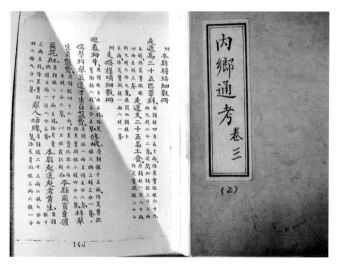

《内乡通考》关于"打春牛"的记载
Records about "Whipping the Ox" in *Nei Xiang Tong Kao*

每年立春的前一天，知县率众官吏卒，身着公服，高擎仪仗和春字牌到城南先农坛，祭祀春牛和芒神。知县亲自扶犁耕地一垅，表示代御亲耕，劝民农桑，以迎春气而兆丰年。然后迎春队伍将土牛、芒神抬至大堂前迎春池旁安放。立春之日，县署仪门大开，对百姓开放，大堂前设香案，摆香烛、祭品。百姓执彩旗，吹唢呐，聚集在大堂前，知县面北而跪，作三献酒，诵读祝文。

On the day before the Beginning of Spring every year, the county magistrate would head his officials and men to go to the Temple of Agriculture in the south of the city,

内乡县衙打春牛
Whipping the Spring Ox at Neixiang County Yamen

in official costumes and with his banners of honor and signs with the character of "spring" held high. There, they would show sacrifices to the spring ox and the God of Spring. The county magistrate would walk behind the ox in plowing along one ridge in the fields, signifying that he was plowing in the place of the emperor and encouraging farmers to be engaged in agriculture and silkworm farming, in order to usher in the spring breath and forebode bumper harvests in the year. Afterwards, the spring-welcoming procession would bring the earthen ox and the god of Spring before the main hall and put them in place beside the spring-welcoming pool. On the day of the Beginning of Spring, the main gate of the County Yamen would be wide open and accessible to all. In front of the main hall would be set up an incense table with incense, candles and sacrificial offerings in place. Ordinary spectators would gather in front of the main hall, carrying colored banners and playing *suona* music. The county magistrate would be kneeling down facing the north, kowtow three times, offer wine, and read aloud the words of toast.

祭祀春牛芒神
Making Sacrificial Offerings to Spring Ox and God of Spring

内乡县衙打春牛
Whipping the Spring Ox at Neixiang County Yamen

礼毕，众官手执彩杖，肃立春牛两旁，赞礼官唱："长官击鼓！"知县三击鼓。赞礼官又唱："鞭春（打春）！"众官吏绕牛三圈，知县将春牛击破。牛肚里事先填满的五谷干果纷纷落地，众官吏与百姓欢呼抢食，以示五谷丰登，吉年有余。随后出县衙开始游街闹春，大街小巷彩棚栉比，披红戴绿。鼓乐、狮子、旱船、高跷等民间杂耍奇玩尽兴表演，万家空巷，官民同乐，到处洋溢着吉庆欢快的气氛，将打春这一活动推向高潮。知县、县丞、主簿视立春日为重大的节日，和家家户户一样吃春盘，饮春酒，拜访长辈。

At the completion of the ritual, the officials would be standing solemnly on both sides of the spring ox with colored sticks held in their hands. The ritual officer would chant, "Sir, beat the drum!" The county magistrate would then beat the drum three times. The ritual officer would then chant, "Whip the spring!" All the officials would walk about the ox for three rounds, and then the county magistrate would thrash into the ox. The five different grains and dry nuts that had been stuffed inside the ox fell onto the ground. The officials and ordinary spectators would happily vie with each other to pick these up from the ground and probably start to eat some of these, which symbolized a bumper harvest of the five different grains in this auspicious year. Subsequently, from the county yamen, the spring-celebrating parade would start, and colorful tents and shades were put up. All kinds of folk performances and variety shows were staged on, including drum beating and other music, lion dances, land boat dancing, and stilt walking. They were so entertaining that people enjoyed themselves tremendously. In all private homes and on streets and in alleys, officials and ordinary people had fun together. Everywhere one went, it was full of joy and auspiciousness and the spring-whipping event was brought to a climax. The county magistrate, his deputy, and the registrar regarded the day of Beginning of Spring as a significant festival, and as everyone in every household, they also had a good bite from the spring plate of snacks, drank spring wine, and visited their elders

内乡县衙打春牛
Whipping the Spring Ox at Neixiang County Yamen

知县击鼓
County Magistrate Sounding the Drum

知县诵读祝文
County Magistrate Reading out the Eulogy

打春牛
Whipping the Spring Ox

百姓抢食五谷干果
People Grabbing Five Cereals and Dried Nuts

内乡县衙打春牛
Whipping the Spring Ox at Neixiang County Yamen

知县扶犁
County Magistrate Working the Plough

据说，知县扶犁亲耕所选用的耕牛和地块，为户主的一大荣幸，还可享受免去当年赋税的优惠待遇。

It was said that it would be a special honor for the farmer if his ox and field had been selected for the magistrate to use in the symbolic and ritual plowing. The farmer would be exempt from the year's taxation as a special favor from the yamen.

据民国《内乡县志》记载，"打春牛"的活动，延续到民国十年（1921年），不过这时的土牛已不是土制，而是纸糊。随着时代的发展，牛耕渐被替代，"打春牛"的岁时活动却逐渐演变成一项民俗文化活动，积淀成为一种文化。每年此时，城乡百姓齐聚、官民同乐，内乡县城内载歌载舞、踩高跷、划旱船、锣鼓唢呐、舞龙舞狮等民俗表演热闹非凡，成为内乡最重要的文化活动之一。

内乡县衙打春牛
Whipping the Spring Ox at Neixiang County Yamen

According to *Nei Xiang Xian Zhi* (Annals of Neixiang County), the annual ritual event of "whipping the spring ox" lasted till the 10th year of the Republic of China period (1921). But by this time, the ox was no longer earthen but was made of paper. In time, the ox would be replaced for plowing the field, and yet the annual event of "whipping the spring ox" evolved into a folk cultural event and into a culture on its own. At this time every year, the ordinary people in towns and villages would gather together; officials and ordinary people would have fun together; inside the county singing and dancing performances were put on; folk performances like stilt-walking, land boat dancing, drum and gong and suona (Chinese trumpet) music, and dragon and lion dances were noisy and inspiring, becoming one of the most important cultural activities in Neixiang County.

丰富多彩的民俗活动
Colorful Folk Activities

相关设施

Existing Artifacts and Facilities

（一）迎春池 Spring-Welcoming Pool

迎春池是内乡县1989年在修复县丞衙挖地基时发现的。它的发现，为研究清代衙门文化、知县职能、民风民俗提供了珍贵的实物依据，具有较高的历史价值、文物价值。

The "spring-welcoming pool" in the Neixiang County Yamen was excavated in 1989 when digging was carried out for the foundation of the project of the county registrar office restoration. The excavation provided valuable artifacts in the study of yamen culture, magistrate functions, and folk customs and culture in the Qing Dynasty, enjoying a very high historical value and cultural relic value.

迎春池为旧时县衙举办岁时活动"打春牛"而凿的石槽。在内乡县衙大堂东山墙外,陈展一大型石槽,名为迎春池。该石槽为长方形,外壁通长136厘米,宽82厘米,高48厘米;内口长102厘米,宽58厘米,深37厘米,可容清水0.22米^3。立春前一日,将土牛芒神移于池旁,打开池盖,里面盛满清水,以祈求风调雨顺。知县在大堂前举行迎春动员后,进行鞭牛,土牛肚内填满的干果散落在地,官吏百姓欢呼抢食,迎春气兆丰年。

迎春池
Spring-Welcoming Pool

The spring-welcoming pool is a stone trough chiseled in the past when the annual ritual of "whipping the spring ox" was carried out at the county yamen. Outside the eastern wall of the main hall of the Neixiang County Yamen, a large stone trough, named "spring-welcoming pool", is displayed. The trough is a rectangle, 136cm long on the outside, 82cm wide, 48cm tall; on the inside, it is 102cm long, 58cm wide and 37cm deep, with a capacity of 0.22m^3 of clear water. One day before the Beginning of Spring, the earthen ox and the God of Spring figures were moved near the pool. The cover of the pool was removed, and it was filled with clear water to the rim, to pray for a year of favorable wind and rainfalls. After the country magistrate carried out the mobilization for the spring-welcoming ritual event in front of his main office hall, he would whip the ox, and the dry fruit and nuts stuffed inside the ox would fall to the ground. Officials and ordinary people would rush to pick them up in competition with each other, in order to welcome the breath of spring in anticipation of a year of bumper harvest.

（二）春凳 Spring Bench

长五六尺，宽逾两尺，可坐三五人，亦可睡卧，以代小榻，或陈置器物，功同桌案。古时民间女儿出嫁，上贴喜花，并置被褥，请人抬着送进夫家。春凳也可供婴儿睡觉使用，故旧制常与床同高。在打春牛中用于摆放祭品。

春凳
Spring Bench

It is five or six *chi* long and over two *chi* wide, accommodating three to five people to be seated. It can also be used as a bed for taking naps, or as a desk or table on which to display artifacts. In ancient times, when an ordinary father married off his daughter, he would have paper-cut characters of "happiness" pasted on it, bedding placed on it, and carried into the home of the son-in-law. The spring bench could also be used as a bed for a baby, so in the old times it was made as tall as a bed. It was used to display sacrificial offerings in the spring ox ritual event.

(三) 春鞭 Spring Whip

春鞭一般用驱使牲畜的鞭子代替，鞭子由鞭杆和鞭条两部分组成。制作鞭杆的小木棍长半米左右，拇指粗细，质地结实。鞭条一般用柔软的动物皮裁割成1厘米宽窄，然后用双股拧结而成。春鞭在打春牛活动中作为鞭打春牛的道具使用。

The whip for driving domestic animals was usually used as the spring whip, which was made up of two sections: the rod and the lash. The rod was composed of a small solid wooden stick of half a meter in length and of a thumb in thickness. The lash was usually made of two kinked strands of animal leather tailored of one centimeter in breadth. The spring whip was used as a prop in the spring ox whipping ritual event.

春鞭
Spring Whip

（四）堂鼓 Drum in the Yamen Hall

鼓类乐器中形体较大者，多使用椿木、色木、桦木和杨木等制作鼓身，因鼓面较大，鼓皮多使用水牛皮。在鼓身上下蒙以两块面积相同的牛皮而成，平常置于木架上用两个鼓槌演奏。大鼓发音低沉而雄厚，古代多用于报时、祭祀、仪仗或军事。

The large and bulky drum as a musical instrument mostly uses Chinese Mahogany, maple, birch, or poplar wood in the making of the barrel of the drum, and because the surface of the drum is large, the drumhead mostly uses buffalo skin. Over the barrel of the drum are stretched two pieces of oxhide of the same size. Most of the time, the drum is placed on a wooden shelf, and two drum sticks are used for performance. The drum gives low, strong and solid sounds and was mostly used in ancient times for announcement of the time, memorial ceremonies, guard of honor, or military purposes.

堂鼓
Drum in the Yamen Hall

（五）锣 Gongs

锣身呈一个圆形的弧面，四周以锣身的边框固定，用木槌敲击锣身正面的中央部分，产生振动而发音。封建时代知县出巡需要鸣锣开道。锣的响数是根据官职的品级而定，三品官以上的敲七至九锣，四品官以下的敲五至三锣，七品官敲三锣半。所谓的半锣，就是将锣敲响后立即用手捂住，没有回音即可。老百姓听到锣声后，坐着的要站起来，扎白头巾的要取下来，以表示对官员的敬重。在打春牛活动中作为知县出巡的仪仗使用。

锣
The Gong

Gongs have a circular-shaped arc surface, fixed with frames of the gong body. A wooden hammer is used to strike the central section of the gong surface, which vibrates and gives the sound. In feudal times, when the county magistrate left his yamen on a business trip, gongs were sounded to warn people to make way for him. The number of gong hammer beats was determined by the rank of the official. Officials of the third *pin* and above were entitled to 7 to 9 beats, officials of the fourth *pin* and below were entitled to 5 to 3 beats, and officials of the seventh *pin* were entitled to 3.5 beats. The so-called half beat was actually this: when the gong was beaten, it was immediately covered by the hand so that no echo was sounded. When ordinary people heard the gong sounds, those seated had to stand up, and those wearing white headbands had to take them down, to show respect to the official. In the spring ox whipping ritual event, gongs were used by the guard of honor for the county magistrate.

(六) 斗 *Dou* (Chinese bushel)

中国市制容量单位，十合为一升，十升为一斗，十斗为一石。在打春牛活动中作为魁星老人的道具。

This was a Chinese unit of dry measure. Ten *he* made up one *sheng*, ten *sheng* made up one *dou*, and ten *dou* made up one *shi*. It was used as the prop for the Old Man Kuixing (Chinese deity in charge of learning) during the spring ox whipping ritual event.

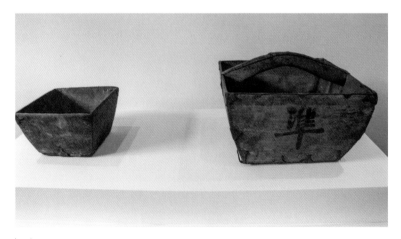

斗
Chinese *Dou*

内乡县衙打春牛
Whipping the Spring Ox at Neixiang County Yamen

（七）官衔牌　Title Signs of the Official

"钦加同知衔、内乡县正五品"官衔牌，木质朱漆、金字。平时插在门口，出行以及举行婚丧礼仪时，拿着前导，以示荣耀。在打春牛活动中作为知县出巡的仪仗使用。

The title signs of "Emperor Appointed Prefecture Deputy Magistrate Rank, Upper Five *Pin* Neixiang County Magistrate" are wooden, painted red, and plated with golden characters. On most days they were firmly upheld at the yamen gate, and they were used to usher the magistrate in front of his procession as symbols of honor, whenever he was dealing with official affairs outside the yamen and when weddings or funerals were held. In the spring ox whipping event, they were used by the guard of honor for the county magistrate.

官衔牌
Title Signs for Officials

（八）肃静、回避牌 "Silence" and "Yield" Signs

木质，白粉底，黑字。官员出行（或者出巡）所乘坐及随行的仪仗，旧时王侯、官员等外出时禁止闲人喧哗之辞。在打春牛活动中作为知县出巡的仪仗使用。

These signs are wooden, having white painted background and black characters, used with the guard of honor for the main official when they were travelling on business (or for inspection), whether the guard of honor was riding or walking. They were used mainly to prevent irrelevant people from making noises when princes, dukes and high-ranking officials were passing by on their business trips. They were used with the guard of honor when the county magistrate travelled to the venue of the spring ox whipping ritual event.

回避牌
A "Yield" Sign

传承保护
Preservation and Protection

非物质文化遗产不是收藏家们锁在柜子里的收藏品，也不是艺术家们孤芳自赏的艺术品，而是中华民族文明赓续、历史文化传承的见证和宝贵财富。内乡县文化文物部门，本着对历史负责、对子孙后代负责的态度，积极对内乡打春牛这一民族文化遗产进行研究挖掘和保护。

内乡县衙打春牛
Whipping the Spring Ox at Neixiang County Yamen

Intangible cultural heritage is not some collection that is hoarded and locked up in the display cabinet, or art pieces that are only enjoyed and appreciated by artists themselves, but the witness for, and valuable assets of, the continuation of Chinese civilization and the inheritance of the Chinese history and culture. The cultural and historical relic departments of the Neixiang County government have conducted active excavation, exploration and protection with regard to this ethnic and cultural heritage, in an effort to be responsible to history and our future generations.

一是挖掘内涵、恢复原貌、再现历史。高度重视"打春牛"的保护工作，拨出专项经费建立机构，组织专人挖掘、整理和保护文化遗产。根据史料记载和老人们的回忆，目前内乡县衙博物馆已恢复了这一习俗。每逢立春日，内乡县衙的"知县"率众"官吏卒"身穿公服，高举仪仗和"春"字牌，到县城郊外扶犁耕田，然后在县衙大堂前举行"鞭春牛"表演，并组织民俗节目展演，成为内乡县一道靓丽的文化景观。2014年，该节目首次走进河南卫视春晚节目，各大媒体高度关注。

First, exploration of the real meaning of the ritual, restoration of the original appearance, and a re-presentation of history. Great attention has been given to the preservation of "whipping the spring ox", and thus exclusive funds have been allocated for the establishment of *ad hoc* organizations, and specifically assigned individuals have been dispatched for the excavation, straightening and protection of cultural relics. This custom has been restored by the Neixiang County Museum, in accordance with historical documentation and the recollections of local old people. Every year on the day of Beginning of Spring, "the county magistrate of the Neixiang County Yamen" would head the large number of officers, officials, and yamen runners, all dressed in official ceremonial costumes, have the guard of honor signs and the "spring" signs, held high, and plow the fields behind the ox in county suburbs. Then the performance of "whipping the spring ox" would be conducted in front of the county yamen main hall. Meanwhile, folk performances would be put

内乡县衙打春牛
Whipping the Spring Ox at Neixiang County Yamen

on for display purposes. Together, these make up the charming cultural landscape of Neixiang County. In 2014, their performance for the first time was presented during the Spring Festival Gala on Henan Satellite Television. This caught the attention of a number of newspapers, journals and online media.

表演现场
Scene of Folk Performances

二是加强保护、加大宣传、形成合力。县衙博物馆积极抓好有关史料搜集、整理、编印、出版、传播工作，加强民族文化遗产的保护和传承。内乡县衙博物馆馆长王晓杰编著的《内乡县衙与打春牛》一书已由中州古籍出版社出版，该书介绍了内乡县衙打春牛这一非物质文化遗产及相关内容，阐述了内乡县衙与清代打春牛活动的历史渊源，图文并茂、形象生动地展现了旧时劝民农桑、迎春打春的传统礼俗，促进了内乡县衙二十四节气民俗文化的保护。同时创新形式和内容，寓教于乐，增强游客互动，调动全社会的力量共同关心支持，形成全民传承、保护、弘扬非物质文化遗产的强大合力。

Second, strengthened protection, increased publicity and integration of all efforts. The county yamen museum has gone all out to organize the collection, editing, compilation, publication and transmission of historical documents in order to enhance the protection and inheritance of the ethnic cultural heritage. The book *Neixiang County Yamen and Whipping of the Spring Ox* by Wang Xiaojie, Director of the Neixiang County Yamen Museum, published and distributed by the Zhongzhou Ancient Books Publishing House, is a work of 220,000 Chinese characters. It has eight chapters, including "historical development of Neixiang County", "Historical Development of Neixiang County Yamen", "Origin and Evolution of the Custom of Welcoming the Spring", "Official Rituals and Customs of Welcoming the Spring in the Qing Dynasty", "A History of the Spring Ox", "Whipping the Spring Ox in the Neixiang County Yamen", and "Other Intangible Cultural Heritage in the Yamen". Vivid descriptions are presented in the book about the intangible cultural heritage of whipping the spring ox in the Neixiang County Yamen and related customs. The book depicts the historical origins of the Neixiang County Yamen and the custom of whipping the spring ox in the Qing Dynasty. The book has abundant illustrations, presenting in a vivid manner of the traditional rituals and customs of encouraging people to do agricultural and silkworm farming, and of welcoming and whipping the spring. This publication promotes the protection of folk customs and cultures of "the twenty-four solar terms" in the Neixiang County Yamen. Meanwhile, new forms and rituals have been innovated to offer fun, education and interaction with tourists. Forces of the whole society have thus been mobilized to help facilitate the cause, and a powerful integrative force has been formed of all the in inheriting, protecting and promoting intangible cultural heritage.

2016年，内乡县《非物质文化遗产——打春牛》陈列展览正式向广大游客开放。以两个展室十二个部分，重点介绍二十四节气、清代官方迎春礼俗、清代民间迎春礼俗、迎春礼的起源与演变、鞭春起源与发展、春牛图、内乡县衙打春牛历史溯源、立春谚语、内乡县衙打春牛掠影、内乡县衙打春牛故事、迎春鸡制作工艺等知识。为了丰富展览效果，还特地制作了扶犁亲耕、东郊迎春、鞭打春牛、游街闹春、古街送春五组泥塑，反映内乡县衙历年来打春牛活动的情景。

In 2016, the Neixiang County Yamen exhibition of "Intangible Cultural Heritage—Whipping the Spring Ox" was officially opened to tourists. It includes 12 sections in two exhibition halls. It focuses on the twenty-four solar terms, official spring, ushering rites and customs of the Qing Dynasty, folk spring ushering rites and customs of the Qing Dynasty, origin and evolution of the spring ushering rites and customs, origin and development of spring ox whipping rites, pictures of spring oxen, historical origins of whipping the spring ox at the Neixiang County Yamen, Beginning of Spring proverbs, glimpses of whipping the spring ox at the Neixiang County Yamen, stories of whipping the spring ox at the Neixiang County Yamen, and the craftsmanship of spring-welcoming roosters. In order to enrich the exhibition effects, five earthen statues have been specifically produced: "Walking Behind the Ox and Plowing in Person", "Welcoming the Spring in the Eastern Suburbs", "Whipping the Spring Ox", "Celebrating the Spring and Parade", and

"Presenting the Spring in Ancient Streets", which portray the scenes of whipping the spring ox at the Neixiang County yamen during all these years.

打春牛陈列展示馆
Display Room of the Whipping the Spring Ox Exhibition

三是主动作为、延续文脉、传播文明。"打春牛"活动 2007 年 6 月被河南省政府公布为第一批省级非物质文化遗产。2014 年 5 月 23 日，由文化部非物质文化遗产司直接领导和中国农业博物馆牵头组织，内乡县衙博物馆与河南省登封市文化馆，湖南省安仁县文化馆、花垣县非遗保护中心等相关单位发起成立了"二十四节气保护工作组"，并联合制定了《二十四节气五年保护计划（2017—2021 年）》，建立并依托二十四节气传习基地，开展相关调查、研究和传承活动。2016年11月30日，"二十四节气"被列入联合国教科文组织"人类非物质文化遗产代表作名录"，内乡县衙博物馆在助力二十四节气申遗、保护传承"立春"节气文化方面做出了相应的贡献。

Third, proactive actions and projects for the continuation of the literary tradition and the transmission of the Chinese civilization. The event of "whipping the spring ox" was designated among the first intangible cultural heritages at the provincial level by the Henan Provincial Government in 2007. On May 23, 2014, under the direct leadership of the Department of Intangible Cultural Heritage, Ministry of Culture, and with the coordination of China Agricultural Museum, the Neixiang County Yamen Museum, together with Dengfeng Municipal Cultural Museum of Henan Province, Anren County Cultural Museum of Human Province, and Huayuan County Intangible Cultural Heritage Protection Center of Hunan Province, launched the "working group for the protection of the Twenty-four solar terms customs", and jointly formulated the "Five Year Protection Plan on the Twenty-Four Solar Terms (2017—2021)", and has been conducting surveys, research and preservation and continuation of the solar-term related customs based on the bases of preservation and study of the Twenty-Four Solar Terms they established together. On Nov. 30, 2016, the "Twenty-Four Solar Terms" was included in the List of "Masterpieces of the Oral and Intangible Heritage of Humanity". The Neixiang County Yamen Museum made its due contribution in facilitating the "Twenty-Four Solar Terms" in applying for the heritage listing and in protecting and promoting the solar term culture of "Beginning of Spring".

图书在版编目（CIP）数据

内乡县衙打春牛／中国农业博物馆组编．—北京：中国农业出版社，2019.5

（"人类非物质文化遗产代表作——二十四节气"科普丛书）

ISBN 978-7-109-25395-7

Ⅰ.①内… Ⅱ.①中… Ⅲ.①二十四节气-风俗习惯-内乡县-通俗读物 Ⅳ.①P462-49②K892.18-49

中国版本图书馆CIP数据核字（2019）第061009号

中国农业出版社出版

（北京市朝阳区麦子店街18号楼）

（邮政编码100125）

责任编辑　张德君　李　晶　司雪飞

北京中科印刷有限公司印刷　新华书店北京发行所发行
2019年6月第1版　2019年6月北京第1次印刷

开本：787mm×1092mm　1/16　印张：3.75
字数：75千字
定价：38.00元

（凡本版图书出现印刷、装订错误，请向出版社发行部调换）